An Unfinished Manuscript of the Scientific Age

Miguel A. Sanchez-Rey

Table of Contents

Royalty in the Scientific Age

5-8

Delirium of the Sciences

10-17

An Autobiographical Study On a Woman's Chair

19-22

PHPR top-scientists and the Feynman Era

24-43

The Guiding Principle

45-47

Royalty in the Scientific Age

The Leading Professor Miguel Angel Sanchez-Rey [*The Grandmaster, The Master of Space-Time*]

The Academy of Advance Science and the Technological Sciences

"Royalty in the Scientific Age has a strange connotation to it, doesn't it? A connotation of belligerence and cruelty. So silence, the red queen!!! The common democratic lineage is -- in many respects, a lineage of political anarchism. Not of a royal bloodline that has autocratic ownership of the religious state."

"Even then, political anarchism (being a distrustful crime) is beholden by a meritocratic monarchy that shows no empathy to the masses. Yet they strive for a global monarchy -- that is the red queen's wishful dream. But a dream of a desecrated monarchal institution. For royalty has no role to play in the Scientific Age, and so silence the monarchs!!! They are heads of state. They do exactly what they're told to do or their guillotine."

"The founding of the superstate is only two hundred years away. But with the permanent decline of the religious state (a crime to be reckon with), the red queen's demise is as imminent as the demise of the world state. And so silence, the red queen!!! For the monarchy has no role to play in the affairs of the superstate."

"The council of the superstate oversees what's left of the permanently declining religious state (until the superstate is quietly dismantled). And that the council of the superstate take authoritative action to both uphold the democratic order and to protect the democratic workforce -- from any internal and/or external threats to its longevity and tranquility. And so silence, the red queen."

"For the federation of democratic industries does not mean wealth and privilege, elitism and bigotry, and/or racism and genocide. Instead, it's over for the royal bloodline. And there's no going back. They brought this upon themselves and they pay the price, and it's not a price to be paid so easily."

"But yet it's not easier said than done. It's a belligerent crime of extreme delirium. And that's the end of it."

Delirium of the Sciences

The Leading Professor Miguel Angel Sanchez-Rey [*The Grandmaster*, *The Master of Space-Time*]

The Academy of Advance Science and the Technological Sciences

Radical science is the horrific response to a breakdown between the avant-garde and the economic establishment during the eve of the Brexit -- the fall of the European nation-state. A vehement criminal act instigated by one of the most frightening crime rings in world history. Whereas Theresa May suspects that the United Kingdom should assert its independence from the European common market, the European state foresees the rising threat of ultra-nationalist hostility. Ultra-nationalism that morph into a neo-Fascist alliance.

An alliance that has radicalized the political establishment and impose ultra-nationalism as its core national policy. Even then -- with the Canadian and the American agreement for a modification to the NAFTA trade deal (proposed by the U.S. President Donald Trump and the Canadian Prime Minister Justin Trudeau), only encourages greater isolationism and nationalist sentimentality -- in the formulation of stipulations and penalties design to discourage global trade and to increase independent economic productivity.

While curtailing immigration policy in the U.S. (which not only helps migrants but also it's largely service base economy) and deregulating key market forces -- that keeps supply and/or demand stable by way of banking regulation. That said, an anticipatory shock in the global economy (beginning with the Asian American trade dispute) has shown that social democracy has reach its permanent demise. Instigated by almost 12 years of irresponsible and atypical economic decision-making (a consequentiality of the 2007-2012 economic depression which spark the emergence of political radicalism).

But the British government's insistence to breakup from the European nation-state -- is nonetheless, a wishful desire to impose the Commonwealth of Nations as the dominant economic union. A union of nations that surrounds the United States.

A growing union seeking to establish strong ties to the British state but in which the North Atlantic Treaty Organization (NATO) is only a few steps away. And while the British monarchy continues to promote the British state, racial supremacy only ensues and a crime of passion becomes a fatal decision to invigorate a global monarchical institution in the form of a powerful dynasty class.

That alone, means that Queen Elizabeth II continues to encourage British neo-Fascism and to provoke the New Left. Assuming that provocation and incitement will propel the solidification of British nationalism. Yet blinded by the factual nature that the European aristocracy could potentially retaliate against the British state to protect what's left of the permanently declining European nation-state.

The queen's wishful dream is the signification of a vehement British prime minister that knows no proper bounds with the European state. And in which scientific radicalism has shown itself to be a deadly force that has cause the political and the scientific establishment, in the U.S. and the U.K., to become ineffective. Ineffective in its unwillingness to neutralize the radicalization of the sciences -- but also in its toleration of scientific radicalism.

The United States has undergone a breakdown in the union of states. And in which the corporate governorship has been solidified (due to a power-struggle between varying competing interests). Resulting in economic policies that favor the wealthier class -- transferring massive wealth, power and privilege to a corporate authority.

Where the constitutional order has since then collapse (a violent breakdown of the centrist-government). And also where American political radicalism has become the center-stage in the U.S. media and tabloids.

The U.S. Congressional elections is only then the foreshadowing of a scientific dictatorship. Whereby a debating body is nonetheless indicative of a failed policy era of cult-policy making.

A debating body that has become a permanent fixture of a broken U.S. constitutional order in which the dominant corporate governorship dictates terms to the free market. Yet the outcry of social activists that are suspicious of the Commonwealth Charter -- it's dire emphases for large concentrations of wealth to a privilege few and the promotion of privilege (in the manner of needless honorship).

Sidestepping the democratic principles of parliamentary democracy stipulated by the Magna Carta.

The radicalization of the sciences -- in its most delirious formulation, has meant that the sciences has falling down an incline plane into both radicalism and fantasy (in which the sciences have lost contact with the reality of the natural and/or social sciences -- the impartiality of the experimental process).

An indication that fame and/or fortune is falsely seen as the driving force of the sciences. But in which war-crime becomes the end consequentiality of such fame and/or fortune.

Where unreason is a deadly guide -- that signifies a frightening Virgil of European and American malevolence. Manifesting as atypical decision-making (of a modernist dystopian reality that will last for *at least half a century*). Understood as a vehement passionate act of both political extremism and the perplexity of scientific radicalization.

And yet a belligerent crime of extreme delirium.

An Autobiographical Study On a Woman's Chair

The Leading Professor Miguel Angel Sanchez-Rey [*The Grandmaster, The Master of Space-Time*]

The Academy of Advance Science and the Technological Sciences

The woman's chair is held in recognition of the highest expectation of achievement and accomplishment in the sciences. But how far is a woman of science willing to go before the chair becomes a chattered professorship? Where excellence and achievement instead becomes years of mismanagement and declining credibility. Yet women are held at a double standard in the academic hierarchy; that much said is true.

But so few woman of science reach the lime light. And so fame is burdensome. It carries its own risk and rewards. And with more and more risks, the greater the rewards. But rewards are often misconstrued as fame and fortune. Whereas fame and fortune are not what drives the sciences, rather it's the commitment for excellence and achievement for its own sake.

And yet years of excellence and achievement does not qualify an expert academic into earning the woman's chair, rather it's the capability to express credibility and commitment to fair, equal and higher standards. A woman's chair is not a chair to be taking lightly but to be a promising beacon of a woman's mentorship and leadership in the sciences.

Whether or not a woman's chair is to be put at a double standard than most chairs is debatable amongst many decision-makers.

But decisions and decisions are to be made that makes the woman's chair a high-stakes enterprise in a planet where most chairs are neutral about sex, gender, and race. But in which all chairs share the same qualities: a commitment to excel as the chair and to surpass the detestability of the chair.

Looks can be deceiving, is the defining lesson of the woman's chair. And so leading by example is not what defines the integrity of a woman's chair, rather it's the capacity to take the leading role in the decision-making of their prospective academic field and/or organization -- in which, the chair is held to the highest standards.

The leading role in which students and/or interns will be able to distinguish between themselves and the detestability of the chair. Whereas the chair brings distinction to their prospective colleges and/or organizations, their students and/or interns will surpass said expectation to accomplish qualifications that surpasses the woman's chair.

PHPR top-scientists and the Richard Feynman Era

The Leading Professor Miguel Angel Sanchez-Rey [*The Grandmaster*, *The Master of Space-Time*]

The Physicalist Program

The Academy of Advance Science and the Technological Sciences

The late Richard P. Feynman is considered the first modern public intellectual in theoretical physics. But as Albert Einstein advocated for a Jewish state or nuclear disarmament, Richard Feynman took part in the Manhattan Project alongside other nuclear physicists -- such as, Enrico Fermi and John Robert Oppenheimer.

Ushering the nuclear era: with the dropping of the first nuclear bomb in Hiroshima and subsequent bombing of Nagasaki three days later (ending the second World War). Richard Feynman eventually left military service to the California Institute of Technology.

Embarking on the development of quantum electrodynamics. Introducing path integrals to further knowledge of quantum behavior in light of Paul Dirac's discovery of anti-particles that set the groundwork for quantum special relativity.

Having won the 1967 Nobel Prize in physics; alongside, Sin Itiro Tomonaga and Julian Schwinger, for advances in quantum electrodynamics, a golden age in particle field theory emerged that brought into play more understanding of the quantum particle interactions.

With Murray Gell-Mann's discovery of quarks that gave way to a deeper understanding of quantum chromodynamics, Stephen Weinberg's and Abdus Salem's completion of the Standard Model -- that unified the electroweak force with the strong and nuclear force:

$$SU(3) \times SU(2) \times U(1)$$

It seem that physics was poised to realize the late Albert Einstein's dream for a grand unified theory of everything that merges all the fundamental forces of nature; including the weakest of all known forces: gravitation.

Yet other physicists were also poised to make more theoretical breakthroughs; such as: supersymmetry and supergravity; most from the European Union (Gerard't Hooft [for advancing knowledge of quantum structure of the electroweak force] and Daniel Z. Freedman [for his discovery of supergravity]). And yet even others, like John von Neumann, made substantial contributions to quantum theory, and to the unraveling of quantum computational methods, that could accelerate progress in achieving artificial intelligence.

Richard Phillips Feynman was distinct to all other physicists of what is consider the heydays of modern physics. Whereas Carl Sagan presented himself as the leading advocate and public intellectual in astronomy and humanist philosophy.

Richard Feynman took part in expert decision-making while also presenting himself as a trusted character that could genuinely reach out to the general public. As a leading educator, encourage the field of physics to become more assessible and worthwhile to the layman of everyday life -- in the form, of lectures and popular science writing.

Physics was poised to make unending breakthroughs, and yet Richard Feynman remain -- in his later life, skeptical of string theory. More prone to a self-referential view of cosmology. Head strong about the importance of the experimental sciences, Richard P. Feynman can be seen as the beginning of a new era where the general public express a strong, skeptical and enduring interest in physics.

Where expertise is an important quality, but yet education remains an important aspect of providing open access to the general public. As citizen scientists begin to emerge. Providing answers and solutions to overlook scientific questions and problems by playing an active part in the sciences. Promoting competition and cooperation in the sciences through public education.

And by proving that world leadership and renowned expertise amongst physicists is an avenue to bring the sciences to a wider readership and audience. Motivating the general public to go along with the hard sciences.

Laying the bases for public intellectualism on a world-wide scale. Where national laboratories collaborate with the public intellectual community to advocate for more funding and notoriety. Yet bring many more scientists together to embark on a world-changing project through enormous governmental and intergovernmental funding, i.e., Los Alamos National Laboratory in New Mexico and CERN at Switzerland.

Excitement was the driving impetus of the Richard Feynman era, but somewhere along the line, public intellectualism fell apart. Made clearer with the catastrophic consequences brought out by the fringe elements of the New Atheist movement and at the extremity of the tea-party (that sought to bring social change through science and/or social justice).

With advocacy for the sciences and the desire for social change, a toxic mix led to the radicalization of the sciences. Eventually the breakdown at CERN, the compromise at ITER, and the havoc at LIGOS, that gave way to the beginning of the Scientific Age. Spawned by the internet, that allowed cult-figured academics to take advantage of their small following to cry havoc on the sciences -- in order, to accelerate their careers in their prospective fields. Even the social justice and anti-war movement cried havoc on the sciences to pressure the political establishment to resolve their social grievances.

Which brought into bear the failings of public intellectualism in the sciences -- that started with the late Albert Einstein. But yet as Albert Einstein sought to discipline the early world order, Richard Feynman instead sought to bridge the publics relationship with scientific experts. Yet others saw a weakness, took advantage of that weakness and nearly fractured Richard Feynman's visionary aim of citizen science and global scientific expertise.

As close to the final theory, the theoretical sciences were met with a severe gap in the publics willingness to trust scientific experts. Succumbing to a modernist dystopian reality, the theoretical sciences went into violent conflict with public intellectualism.

In addition to the private sector battling the government sector, public intellectualism collapse and Richard Feynman's legacy remain an uncertain predicament to much of the expert leadership in international academia. How else is the public not to be frightened of the sciences, if the expert leadership in the sciences (understood as the intelligentsia) cannot address the sciences to the general public?

How else is skepticism and the experimental method be conveyed, if it risk to incite radicalization and profound political differences? These are problematics that cause serious havoc to many leading academics (for example: not being able to take field trips with the military industrial complex, least the chain of command is to breakdown due to deep moral differences with the anti-war movement).

It seem the careers of many public intellectuals, that held on to an extreme form of public intellectualism, came to an end. Instead, cult-figured, their followers disenchantment with their unending struggle for fame and fortune, meant that even the private sector retaliated.

That is, the radicalization of the sciences spelled the end of modern public intellectualism and a desire to return to the soft nature of the Feynman era. An era that advocated for scientific progress but remain far separated from long-term decision-making.

Where expert scientists could convene a meeting with politicians. And politicians would weigh their options to go along with these expert scientists to pursue further scientific research (base on public opinion and/or national defense). Very much understood as the driving force for progress in the sciences during the height of the Cold War. Propelling not only progress in high-energy physics but also in the technological sciences.

Yet remain far apart from decision-making in economics or far away from environmental activism. And that the public not encroach into the sciences but to see the expert-scientist as a beacon that will lead to a genuine career in the core sciences.

While others dwell on other important matters of social justice and war theory, as experts in the philosophical and political sciences, expert scientists remain dissociated from political and ideological interests.

That public intellectualism went the wrong way with Carl Sagan's visionary aim of advocacy of scientific progress through social change. Encouraging contemporary public intellectuals to overstep themselves in the sciences. Causing havoc to the non-impartiality of both scientific decision-making and peer-review.

Eventually with open access -- encouraging radicalism and extremism, allowing the 2017-2018 Nobel Prizes to fall victim to neo-Fascism and an unprecedented crime ring of atypical decision-making. Even Neil de Grasse Tyson presented a difficult dilemma of false advocacy of social change and the sciences -- in an attempt, to reinvigorate his grip on fame (with the publication of *Accessory to War* and his advocacy of Space Force without any genuine qualification of what it means to save and/or take a life in the battle field, or even any experience of the drawbacks and pitfalls of the U.S. war machine, in an era of Republican tea-party politics).

The incompetence of Carl Sagan's visionary aim has left a haunting message to many expert scientists. While The Physicalist Program [PHPR] top-scientists are task to complete a set task, they are consider to be renowned expert decision-makers and world leaders in the sciences.

Whom can bridge the gap between competent decision-making in the sciences and the capacity to relate to the public -- in a responsible and clear-cut manner.

While pursuing the high-stakes task, they present themselves in all secrecy -- in their early adult lives, but live their later adult lives far apart from the general public. But are seen as role models and revered figures that can forestall the unforeseeable catastrophic scenario (acknowledge as geniuses at crises control and game theory).

Superior to any decision-maker and any world-leader in any particular field, are capable of disciplining the public to trust the sciences while encouraging the general public to pursue the sciences. Yet not forgetting that -- though the Scientific Age is an age of wild anticipation, the citizen scientists has a large role to play.

While contemporary public intellectuals fragment the sciences, citizen scientists bring a genuine unified force in the sciences that can contribute to the scientific process that eventually leads to significant progress in the Scientific Age.

Yet have a responsible viewpoint not to embed their own preconceptions and biases into the blind-review process (and have a strong attachment to the experimental method). Even be more than qualified to take part in the sciences -- in anticipation, of taking part in both PHPR and the Scientific Age.

Be seen as role models that can take control of scientific leadership and decision-making. Becoming world-leaders and renowned experts that surpasses the limits of public intellectualism.

Understood as a return to the Richard Feynman era of citizen science and scientific intellectualism in the Scientific Age. Whereby the PHPR top-scientists hold an undying commitment to the experimental and theoretical sciences. An oath to public service. And yet have a wildly strong attachment to public participation in the sciences.

The Guiding Principle

The Leading Professor Miguel Angel Sanchez-Rey [*The Grandmaster, The Master of Space-Time*]

PHPR

The Physicalist Program [PHPR] is design to save lives (as a resolution to a foreseeable catastrophic scenario in the Scientific Age) in the form of a task. *The Grandmaster* is to complete a task and set the next task.

The First Task of PHPR is a 100 Year Task.

The stringent task is to be set by Trinity College, University of Cambridge.

Selection into PHPR is to be decided by all intelligence agencies under jurisdiction of NATO (North Atlantic Treaty Organization).

PHPR top-scientists are task to further PHPR's directive. Stating that the top-scientists in PHPR is to take whatever action is necessary to protect and to complete the set task -- so that, there is no internal and/or external interference.

PHPR's task is not to be fully disclosed until completion is in the near horizon.

When lives are at stake, the top-prize is of little concern.

The Master of Space-Time is to forfeit The Master of Space-Time in return for The Leading Professorship after The PHPR Protocol is completed.

Otherwise, *The Master of Space-Time* is to be earned through a significant achievement in the Scientific Age. Such an achievement must be decided by a unanimous and joint decision from NATO and UNESCO (United Nations Educational, Scientific and Cultural Organization) -- in which, *The Master of Space-Time* is to remain classified until full retirement from their prospective academic institution and/or learn academy. After which, *The Master of Space-Time* is to be decided by a binding decision from the council of the superstate.

PHPR is to be dismantled when the last task is completed. By then the Scientific Age has come to an end…

The AASTS

The Academy of Advance Science and the Technological Sciences [AASTS] mandate is to do the right thing -- when giving the order to do the right thing.

The utmost requirements are set by the AASTS Protocol not far from Sorbonne University, France.

When giving the order to do the right thing, the leading scholar is to take leadership position rather than the top-prize. While The Leading Professor will continue to further higher learning.

The fellowship and all other learn academies are to respect the boundaries between The Leading Professorship and the learn academy. And to resort to indirect channels to establish (only short-term) diplomatic relations with The Leading Professorship -- in only, the most extreme scenario.

After retirement from the AASTS, the leading professorship is forfeited for *The Master of Space-Time*. *The Grandmaster* retains the leading professorship as a life-long honorific.

Yet *The Grandmaster* will continue on to ascertain and establish the Advance Age.

The AASTS Protocol is to be reset by *The Grandmaster* at the dawn of the Advance Age. While the remaining academic hierarchy is to then be quietly dismantled -- in anticipation, of achieving academic harmony and the founding of the federation of democratic industries. Accomplishing the complete mastery of space-time (and of higher qualification), *The Grandmaster* is the master of its own program, academy and futility. But the founder set The First Task. Then laid the foundation for an academy of rational existentialism after the collapse of the magisterium. And is declared *the Atheist Saint* and *The Saint of the Advance Age* -- the guiding principle of the completion of a task that led to The First Task.

The AASTS is to be dismantled by a unanimous agreement and official declaration at the beginning of the incalculable technological stage.

www.ingramcontent.com/pod-product-compliance
Lightning Source LLC
Chambersburg PA
CBHW062343220526

45469CB00008B/2813